D0318951

V
EA

&
AKE

Skull from Herculaneum

Lava bomb

Sulphur

Gray-Milne seismograph, 1885

Carbonized bread from Pompeii

Cut peridot

Gem-quality olivine

Cut and uncut diamond

Carbonized walnuts from Pompeii

Preserved eggs from Pompeii

Body cast from Pompeii

Pele's hair

Voyager 1
space probe

# EYEWITNESS
# VOLCANO &
# EARTHQUAKE

Written by
SUSANNA VAN ROSE

Bottle melted in
eruption of
Mount Pelée

Perfume bottle
melted in eruption
of Mount Pelée

Zhang Heng's
earthquake detector

DK

Seneca, Roman philosopher who wrote about earthquake of 62 CE

Title page from *Campi Phlegraei*

Fork and pocket watch damaged in eruption of Mount Pelée

![DK](DK logo)

LONDON, NEW YORK,
MELBOURNE, MUNICH, and DELHI

**Project editor** Scott Steedman
**Art editor** Christian Sévigny
**Designer** Yaël Freudmann
**Managing editor** Helen Parker
**Managing art editor** Julia Harris
**Production** Louise Barratt
**Picture research** Kathy Lockley
**Special photography** James Stevenson
**Editorial consultants** Professor John Guest and Dr Robin Adams

RELAUNCH EDITION (DK UK)
**Editor** Ashwin Khurana
**Managing editor** Gareth Jones
**Managing art editor** Philip Letsu
**Publisher** Andrew Macintyre
**Producer, pre-production** Nikoleta Parasaki
**Senior producer** Charlotte Cade
**Jacket editor** Maud Whatley
**Jacket designer** Laura Brim
**Jacket design development manager** Sophia MTT
**Publishing director** Jonathan Metcalf
**Associate publishing director** Liz Wheeler
**Art director** Phil Ormerod

RELAUNCH EDITION (DK INDIA)
**Senior editor** Neha Gupta
**Senior art editor** Ranjita Bhattacharji
**Senior DTP designer** Harish Aggarwal
**DTP designer** Pawan Kumar
**Managing editor** Alka Thakur Hazarika
**Managing art editor** Romi Chakraborty
**CTS manager** Balwant Singh
**Jacket editorial manager** Saloni Singh
**Jacket designers** Suhita Dharamjit, Dhirendra Singh

This Eyewitness ® Guide has been conceived by
Dorling Kindersley Limited and Editions Gallimard

First published in Great Britain in 1992.
This relaunch edition published in 2014
by Dorling Kindersley Limited, 80 Strand, London WC2R 0RL

Colour reproduction by Alta Image Ltd, London, UK
Printed and bound by South China Printing Co Ltd, China

Discover more at
**www.dk.com**

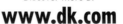

Mining transit

Lava stalagmite

Figure of Zhang Heng,
Chinese seismologist

# Contents

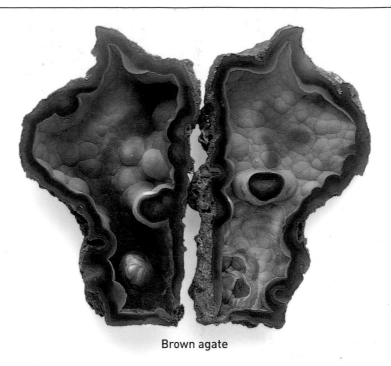

Brown agate

# An unstable Earth

Volcanoes and earthquakes are nature run wild. An erupting volcano may bleed rivers of red-hot lava or spew great clouds of ash and gas into the sky. During a severe earthquake, the ground can shake so violently that entire cities are reduced to rubble. These natural disasters can be terrifying. But most eruptions and earthquakes cause little damage to people or property.

**The perfect volcano**
The graceful slopes of Mount Fujiyama in Japan are shown in this print by Katsushika Hokusai (1760–1849). This dormant (sleeping) volcano (pp. 38–39) is an almost perfect cone.

**Wall painting**
Nearly 8,000 years old, this wall painting is the earliest known picture of a volcano. It shows an eruption of the Hasan Dag volcano in Turkey.

### Ash treatment
Volcanic eruptions can have benefits. In Japan, being buried in warm volcanic ash is thought to be good for the health.

### Ashy volcano
Observing ashy volcanic eruptions (pp. 14–15) from the ground can be dangerous. This image of the Augustine volcano in Alaska was taken from the safety of a satellite. The ash cloud is being blasted 11 km (7 miles) high.

### Old Faithful
Geysers spit boiling water high into the air (pp. 36–37). Old Faithful, an American geyser, has erupted every hour for at least 100 years.

### Back from the dead
This painting is of a boy killed during a quake in Assisi, Italy. Legend has it that St Francis of Assisi brought the boy back to life.

### Spitting fire
Mount Etna rises 3,390 m (11,122 ft) over the Italian island of Sicily, and is one of the highest and most active volcanoes in Europe. Fountains of gassy lava often spew from the summit (left). The nearby town of Catania is occasionally showered with ash from explosions.

### San Francisco, 1989
In 1906, San Francisco was flattened by an enormous earthquake. Earthquakes of this size seem to rock the area every hundred years or so. A smaller quake on 17 October, 1989, shook many waterfront houses off their foundations. Some 62 people died in the 15 seconds of shaking.

# Fire from below

Deep inside the Earth, rocks melt into a thick liquid called magma (molten rock). Most of the molten rock spewed out by volcanoes comes from the top of the mantle – a layer that lies between the Earth's crust and its outer core. Because magma is hotter and lighter than the surrounding rocks, it rises, melting some of the rocks it passes on the way. If it manages to find a way to the surface, the magma will erupt as lava.

### Hot as hell
In the Christian religion, hell is described as a fiery underworld, as shown in this 1788 painting by Irish artist James Barry.

### Channels of fire
In this 17th-century engraving, Athanasius Kircher imagined that Earth had a fiery core which fed all the volcanoes on the surface. We now know that little of the planet's interior is liquid.

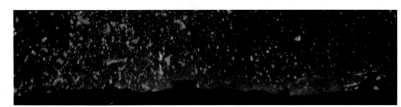

Red-hot lava (liquid rock) shoots out of a volcano in a curtain of fire

### Into the crater
Jules Verne's famous story *Journey to the Centre of the Earth* begins with a perilous descent into the crater of Mount Etna.

### Layers of Earth
Just like an apple, the Earth is made up of layers. Its core is surrounded by a moving layer of rock called the mantle. The hard, outer layer is known as the crust.

### Basalt
The ocean floors that cover three-quarters of the Earth's surface are made of a dark, volcanic rock called basalt.

### Granite
The continents are made of rocks that are similar to granite – a lighter rock than basalt.

### Iron hot

Pure iron melts when it reaches 1,535°C (2,795°F). Most of the Earth is hotter than this.

*Lithosphere, which includes tectonic plates (pp. 10–13)*

### Iron heart

The iron meteorites that fall to Earth are thought to be pieces of the cores of planets that have broken up.

*Inner core of solid metal*

*Upper mantle*

*Lower mantle*

*Outer core of liquid metal*

**Ultramafic nodule**

**Ultramafic nodule**

### Inside the Earth

The mantle, lying beneath the Earth's thin, rocky crust, is solid, but it produces pockets of magma that feed volcanoes on the surface. Inside the mantle is the Earth's metal core. This consists of an outer core of liquid metal wrapped around a smaller, solid inner core.

### Inner secrets

No drill hole has yet reached as deep down as the mantle. But, occasionally, rising magma breaks off fragments of the mantle on its way to the surface. Known as ultramafic nodules, these fragments of very heavy mantle rock are found in erupted lava flows.

## Carrying the weight of the world
In Roman mythology, the god Atlas held the sky on his shoulders. In this statue, he carries the entire globe.

# Tectonic plates

Volcanoes and earthquakes are more common in some parts of the world than others. In the 1960s, the secrets of the deep ocean floor began to be revealed. Scientists discovered that the Earth's crust is made up of huge slabs of rock that fit together like odd-shaped paving stones. Called tectonic plates, these chunks of rock move constantly across the surface of the planet at a rate of a few centimetres a year. Most volcanoes and earthquakes occur at plate boundaries where the tectonic plates collide, rub together, or move apart.

## Continental drifter
A German scientist, Alfred Wegener (1880–1930), first used the term "continental drift". He saw the fit of the coastlines of South America and Africa and suggested that they had once been attached.

However, his theory that the continents could have moved apart was largely ignored. His ideas were accepted only when spreading ridges (pp. 24–25) were discovered 40 years later.

## Ring of Fire
The "Ring of Fire" is an area in the Pacific Ocean where most of the world's volcanic and earthquake activity occurs. On this map, the black cones are volcanoes and the red zones are prone to earthquakes.

## Lessons of history

This plaster cast shows a man killed in the eruption of Mount Vesuvius, which devastated the Roman towns of Pompeii and Herculaneum in 79 CE (pp. 26–31).

## Living on the Ring of Fire

Volcanoes and earthquakes are frequent events in Japan. This huge quake in 1925 damaged the ancient city of Kyoto.

*Iceland sits on top of the Mid-Atlantic spreading ridge (pp. 24–25)*

*Like Japan, Kamchatka is part of the Pacific Ring of Fire*

*Alaska and the Aleutian Islands have many volcanoes and earthquakes*

*The Mid-Atlantic Ridge is part of the largest mountain range in the world*

*The island of Réunion was formed by a hot spot (pp. 22–23) that was under India 30 million years ago*

*Indonesia, home to over 125 active volcanoes, is at the boundary of two plates*

*Antarctica is surrounded by new ocean made by spreading ridges (pp. 24–25)*

Mount Erebus, an active volcano in Antarctica

## Drifting plates

This globe has been coloured to highlight the tectonic plates. It is the plates, and not the continents, that are on the move.

*There are no active volcanoes in Australia, which sits in the middle of a plate*

## Fire-land

Iceland is made almost entirely of volcanic rocks like those found on the deep ocean floor. It has gradually built up above sea level through intense and prolonged eruptions.

# Moving plates

Most volcanoes are found at plate boundaries, where melting rock forms columns of magma that erupt at the surface. When two plates move apart, a chain of gentle volcanoes, known as a spreading ridge, is formed. Where plates collide, one is forced beneath the other, forming a subduction zone. The sinking plate partly melts and the hot, liquid magma rises. A third kind of volcano erupts above a hot spot, a place where rising magma burns through the Earth's crust.

*Ocean floor is older the farther it is from the ridge*

*Rift (crack) where ocean floor is pulled apart*

*Old volcanoes that have moved away from hot spot*

*Hot-spot volcano, builds up a mountain so large it forms an island*

50 km (31 miles)

100 km (62 miles)

*Magma erupts in the rift as pulled-apart plates release pressure*

*Hot plume of magma rises to form a hot spot*

*The new plate cools and thickens as it moves away from the heat of the ridge*

## Spreading ridges

New ocean floor is made where plates are pulled apart (pp. 24–25), creating a rift (crack in the Earth's crust). Here, magma erupts as lava, creating new rock. All the ocean floor has been made this way in the last 200 million years.

## Hot spots

Hot spots (pp. 22–23) are areas in the middle of a tectonic plate, where columns of magma from the mantle rise to the surface and punch a hole in the plate, forming a volcano.

## Volcano chain

Guatemala in Central America is home to a chain of volcanoes, many still active. They sit on top of a subduction zone, formed as the Cocos Plate sinks beneath the larger North American Plate.

## America's fault

The San Andreas fault is probably the most famous plate boundary in the world. The plates, which constantly slide against each other, move about 1 cm (0.4 in) a year.

*A deep ocean trench is formed where the ocean floor descends in a subduction zone*

*Continental boundary lifted above subduction zone*

*There are few volcanoes along transcurrent plate margins*

*Ocean plate heated as it plunges into the mantle*

*Lightest melted rock rises through the surrounding dense rocks*

*Magma reservoir feeds volcano*

*Lithosphere (crust and very top of mantle)*

*Asthenosphere (soft, upper part of mantle)*

## Subduction zone

Where one plate is pushed below another, it sinks into the mantle, partly melting some rocks to form magma. This magma erupts at the surface through volcanoes.

## Transcurrent plate margin

Where two plates meet at an odd angle, a boundary called a transcurrent plate margin is formed.

# Mount St Helens

When Mount St Helens, a volcano in the Cascade range in northwest USA, blew its top on 18 May 1980 it had been quiet for 123 years. The massive explosion was heard in Vancouver, Canada, 320 km (200 miles) away. The blast tore off most of the north side of the volcano, leaving a gaping hole. Pyroclastic flows of hot ash and gas (p. 16) rushed down the slopes at terrifying speeds. The explosion continued for nine hours, sending millions of tonnes of ash 22 km (14 miles) up into the atmosphere. Vast areas of forest were flattened and 57 people were killed.

**Slumbering giant**
Before the eruption, Mount St Helens had a beautiful, snow-capped peak.

### 38 seconds after the first explosion
After two months of small explosions, the north slope of Mount St Helens suddenly shivered and seemed to turn to liquid. This picture, taken 38 seconds into the explosion, shows the avalanche roaring down the north face and a cloud of ash and gas blasting skywards.

*Ashy eruption cloud*

*Feeder pipe*

*Reservoir of hot, gassy magma*

### Feeding the fury
Lighter than the solid rock around it, hot magma had risen under Mount St Helens. It had gathered in an underground pool, called the magma reservoir. The hot rock reached the crater along a feeder pipe.

### Four seconds later...
... the avalanche of old rock had been overtaken by the darker, growing cloud of ash, which contained newly erupted material. Gary Rosenquist, who took these pictures from 18 km (11 miles) away, said later that "the sight... was so overwhelming that I became dizzy and had to turn away to keep my balance".

### Moving wall of ash

As the ash cloud blasted out, it became lighter than air and began to rise. Gary Rosenquist took this picture before he ran for his car. "The turbulent cloud loomed behind us as we sped down Road 99," he wrote later. "Mudballs flattened against the windshield. Minutes later it was completely dark."

### Last gasp

In the months after the eruption, thick, sticky lava was squeezed from the magma reservoir like toothpaste from a tube. It formed a bulging dome, which reached a height of 260 m (853 ft) in 1986. The dome later crumbled into lava fragments.

### Flattened forests

Mature forests of trees up to 50 m (164 ft) tall were flattened by the blast of the eruption.

### Eleven seconds later...

... the avalanche of old rock had been completely overtaken by the faster blast of ash. On the right, huge chunks of rock could be seen as they were catapulted out of the cloud.

# Ash and dust

Lapilli, bite-sized fragments of frothy lava

Ash, smaller pyroclastic fragments

Dust, the smallest and lightest lava fragments

Explosive volcanoes pour clouds of ash into the sky. The ash is formed from volcanic rock that has been blown into billions of tiny pieces. These rock fragments, known as pyroclastics, range from huge lava blocks (p. 18) to fine, powdery dust (pp. 34–35). Between these two extremes are lapilli (Latin for "little stones") and ash. Sometimes the ash clouds collapse under their own weight, forming pyroclastic flows or surges. Unlike lava flows, pyroclastic flows can be extremely dangerous.

**Constructing a cone**
Mountains are built up as pyroclastic fragments settle in layers on a volcano's slopes. Gassy fire-fountain eruptions build cinder cones of bombs and ash.

Prehistoric pyroclastic flow deposit, near Naples, Italy

Fine-grained ash

Pumice bomb

Lithic (old lava) fragment

Detail of Neapolitan pyroclastic flow deposit

**Volcano biography**
Frozen in a volcano's slopes is a record of its past eruptions. The rock layers can be dated and their textures analysed. The ash layers in this rock (right) were erupted by an English volcano about 500 million years ago.

**Glowing avalanches**
If the erupted mixture of hot rocks and gas is heavier than air, it may flow downhill at more than 100 kph (60 mph). Such a pyroclastic flow may flatten everything in its path. Equally destructive are pyroclastic surges – flows that contain more hot gas than ash.

# Night of the ash cloud

After lying dormant for 600 years, Mount Pinatubo in the Philippines began erupting in June 1991. Huge clouds of ash were thrown into the air, blocking out the sunlight for days. Over 100 m (330 ft) of ash lay on the upper slopes of the volcano. Torrential rains followed, causing mud flows that swept away roads, bridges, and several villages (p. 56). At least 400 people were killed and another 400,000 were left homeless.

Their fields buried in ash, farmers take their buffaloes and head for greener pastures

## Buried crops
A thin fall of ash fertilizes the soil (pp. 40–41), but too much destroys crops. Whole harvests were lost in the heavy ash falls that followed the eruptions of Mount Pinatubo.

### Breathing easy
Every step raises fine ash that fills the air. Covering the mouth and nose with a wet cloth helps to keep the throat and lungs clear.

# Fiery rocks

Volcanoes erupt red-hot lava. Sometimes the lava oozes gently from a hole in the ground. At other times it is hurled into the air in spectacular fire fountains. When it lands, the lava flows down in rivers of hot rock that can cover the countryside. If the lava is less fluid, explosions may occur as volcanic gas escapes from the hot rock. These explosions throw out chunks of flying lava, known as bombs and blocks.

**Reheated lava**
Some of the gas dissolved in lava is lost when it cools. This piece of lava frothed up when it was reheated, showing that it still contained a lot of its original gas.

Dense, round bomb

Bomb thrown out by Mount Etna on the island of Sicily in Italy (pp. 6–7)

**Bombs and blocks**
Bombs and blocks can be as big as houses or as small as tennis balls. Bombs are usually more rounded, while blocks are more dense and angular. Their shapes depend upon how fluid or gassy the lava was during the eruption.

Small, explosive eruption photographed at night on Mount Etna

**Twisted tail**
The odd twists and tails of many bombs are formed as they spin through the air.

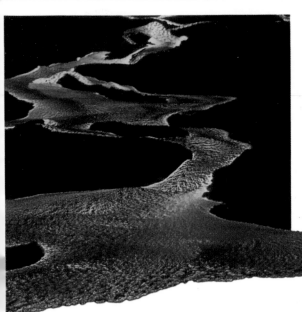

## Hawaiian aa

Glowing red at night, the intense heat of an aa flow shows through the surface crust of cooling lava. Lava flows take a long time to cool. As they harden, the flows grow thicker and slow down.

Hardened chunk of ropy pahoehoe lava

### Pahoehoe flows

Pahoehoe is more fluid than aa and contains more gas. As its surface cools, the flow grows a thin skin. The hot lava on the inside makes the skin wrinkle so its surface looks like the coils of a rope. The crust may grow so thick that people can walk across it while red-hot lava continues to flow in a tunnel below (p. 23).

# Aa and pahoehoe flows

Lava flows pose little danger to people as they rarely travel faster than a few kilometres an hour. The two kinds of flows get their names from Hawaiian words. Aa (pronounced *ah-ah*) flows are covered in sharp, angular chunks of lava known as scoria. Pahoehoe (*pa-hoy-hoy*) flows grow a smooth skin soon after they leave the vent.

Droplets of remelted lava from the roof of a pahoehoe tunnel

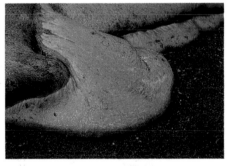

### Pahoehoe toe

This picture shows red-hot pahoehoe bulging through a crack in its own skin. New skin is forming over the bulge. A pahoehoe flow creeps forwards with thousands of little breakouts like this one.

### Spiny and twisted

This chunk of scoria from the surface of an aa flow was twisted as it was carried along.

### Fire and water

Volcanic islands like Hawaii are usually fringed by black beaches. The sand is formed when hot lava hits the sea and is shattered into tiny particles. It is black because the lava is rich in dark minerals.

Black sand from the volcanic island of Santorini, Greece

# Gas and lightning

Captain Haddock and friends flee from a volcano's gases in the Tintin adventure *Flight 714*

Volcanic gases are extremely dangerous. In August 1986, an explosion in Lake Nyos in Cameroon, Africa, released a cloud of volcanic gases. The poisonous fumes killed 1,700 people living in villages nearby. The main killer was carbon dioxide, which has no odour and is very hard to detect. Volcanic gases, on the other hand, are extremely smelly; hydrogen sulphide, for example, smells like rotten cabbage, and the acid gases hydrogen chloride and sulphur dioxide sting the eyes and throat. They also eat through clothes, leaving holes with bleached haloes around them.

**Raising a stink**
Nearly 40 years after the last eruption of Kawah Idjen in Java, Indonesia, sulphur and other gases are still escaping into the volcano's crater.

**Steam-assisted eruption**
When the new island of Surtsey was formed off Iceland in 1963 (p.41), sea water poured into the volcano's vent and hit the hot magma, producing spectacular explosions and huge clouds of steam.

**Gas mask**
Made to protect the wearer against low concentrations of acid gases, this gas mask also keeps out all but the finest volcanic dust.

Volcanologist studying Hawaiian lava flows behind the safety of a gas mask

## Lightning flash

Immense flashes of lightning are often seen during eruptions. They are caused by a build-up of static electricity produced when tiny fragments of lava in an ash cloud rub against each other. This picture shows lightning bolts at Mount Tolbachik in Siberia. It was taken during the day – the Sun can be seen on the left, shining through a cloud of dust and gas.

## Vesuvius

Lord Hamilton, British ambassador to Naples, saw lightning flashes as he watched the 1779 eruption of Vesuvius (p. 31).

## Floating rock

The volcanic rock pumice is light because it is full of bubbles of gas. Some pumice is light enough to float on water.

## Floating on an acid lake

Volcanologists sample volcanic gases from an acid lake in the crater of Kawah Idjen. The gases rising from the volcano are dissolved in the lake water that fills much of the crater. Such acid lakes are very hostile to life, and would burn a swimmer's skin in minutes.

# Hot spots

The largest volcanoes on Earth are found above hot spots – areas deep within the Earth's mantle that produce huge volumes of magma. When the hot magma rises, it burns through the Earth's crust creating a volcano (pp. 12–13). Many volcanic islands, such as Iceland and Hawaii, are located above hot spots. The Hawaiian island chain is part of a huge undersea mountain range that formed over millions of years as the hot spot erupted great volumes of lava onto the moving plate above it.

### Mauna Loa erupts
Two of the world's biggest volcanoes, Mauna Loa and Kilauea, are on the volcanic island of Hawaii. Here, fire fountains erupting from Mauna Loa have created black cinder cones made of ash and bombs (p. 16).

### Volcano goddess
Some Hawaiians believe that the goddess Pele makes mountains, melts rocks, and builds new islands. The fiery goddess is said to live in the crater Halema'uma'u, at the summit of Kilauea.

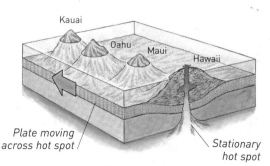

Kauai
Oahu
Maui
Hawaii

*Plate moving across hot spot*

*Stationary hot spot*

### A string of islands
The Pacific Plate is moving over the stationary Hawaiian hot spot, under the south end of the island. In addition to Mauna Loa and Kilauea, a third volcano, Loihi, is growing below the sea to the south. The north end of Hawaii is made up of older, extinct volcanoes.

### Pele's hair
The hot, fluid lava of a Hawaiian fire fountain may be blown into fine, glassy strands. These are known as Pele's hair.

### Moving hot spot
This volcano is Piton de la Fournaise on the island of Réunion in the Indian Ocean (p. 11). The island is the tip of a huge volcano that rises 7 km (4 miles) above the ocean floor. The hot spot has moved 4,000 km (2,500 miles) in the last 30 million years.

Lava has solidified around this tree, leaving a tree mould

Road buried by lava during eruption of Kilauea

### Up in flames
Lava in tubes remains hot and fluid, so it can travel many kilometres, engulfing land and villages on the way.

Lava stalagmite made of drips in a pahoehoe tube

### Lava tube
The skin of a pahoehoe flow (p. 19) may crust over into a roof thick enough to walk on. Only a metre or so below, hot lava continues to run in a tunnel or "tube". Lava dripping off the underside of the roof creates strange formations called lava stalagmites and stalactites.

# Ocean floor

New ocean floor is constantly being made by volcanic eruptions beneath the waters. Spreading ridges form where two plates pull apart, creating a rift (crack) in the ocean floor. Here, the lava erupts gently, forming rounded shapes known as pillow lava. This new rock fills in the widening rift as the plates pull further apart. In this way the oceans grow just a little wider, a centimetre or so, a year. In places, the rifts are bubbling with hot springs, known as black smokers, that are home to a variety of strange lifeforms.

**Rift through Iceland**

The Skaftar rift (above) is part of a 27-km- (17-mile-) long rift that opened in 1783. It erupted 13 km³ (3.1 miles³) of lava over eight months. More than 10,000 Icelanders died in the famine that followed.

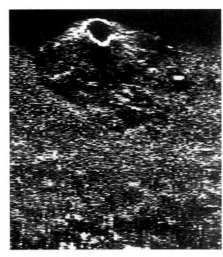

**Undersea volcano**
This marine volcano lies 4,000 m (13,120 ft) below the Pacific Ocean and measures 10 km (6 miles) across.

Icelandic eruptions give a glimpse of how spreading ridges make new oceanic plate. The eruptions tend to be from long cracks, rather than central craters.

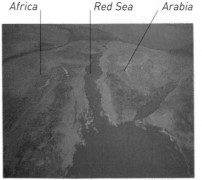

Africa          Red Sea          Arabia

**Red Sea**
A spreading ridge runs through the Red Sea. For the last 20 million years it has been making a new ocean floor, as Arabia moves away from Africa.

**Submersible** *Alvin*, taking photos of ridges

# Black smokers

These strange chimney-like structures are found along spreading ridges on the ocean floor. The black, acidic water that gushes up from these hot springs is rich in valuable metal minerals produced by the new ocean plate that is formed at the ridges.

Sulphur-eating tube worms from the Galapagos rift

Rounded, pillow-shaped underwater lava

## Manganese nodules

The ocean floor is carpeted with black lumps that are rich in manganese and other metals.

## Sea urchins

The animals that live around black smokers include clams and mussels up to 30 cm (1 ft) long. These urchins were seen on the Galapagos rift.

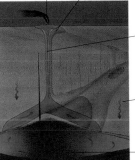

*Plume of black metal minerals*

*Black smoker chimney*

*Feeder channel or pipe*

*Cold sea water seeping through hot rock*

*Magma reservoir*

Model of black smoker

## Lava feeder channels

Two ancient lava feeder channels can be seen in the rock above.

*Coarse crystal structure indicates slow cooling*

## Chimney pipes

When the rising minerals meet the chilled ocean water, they cool and harden to form the chimney pipes that surround the black smokers. These grow steadily, collapsing only when they get too tall.

Gabbro, a coarsely crystalline rock from an old seafloor magma reservoir in Cyprus

25

# Mount Vesuvius

**Pliny the Younger**
This scholar watched the eruption from across the Bay of Naples.

Perhaps the most famous eruption of all time was that of Mount Vesuvius near Naples, Italy. When the long-dormant volcano erupted in 79 CE, the residents of the Roman towns of Pompeii and Herculaneum were caught unawares. Hot ash and stones rained down on Pompeii for hours until it was buried several metres deep. Many trying to flee were overwhelmed by a sudden blast of ash and gas (a pyroclastic surge, p. 16). The buried towns were virtually forgotten until excavations began in the 18th century. These digs have since unearthed priceless treasures.

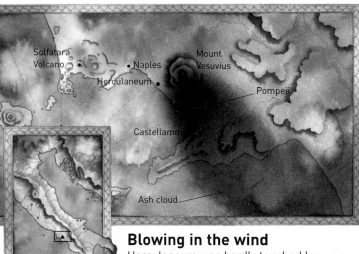

**Blowing in the wind**
Herculaneum was hardly touched by the falling ash that rained on Pompeii. But the pyroclastic flows and surges (p. 16) that followed affected both towns.

**Beware of dog**
This floor mosaic from a Pompeii entrance hall was designed to warn off intruders.

**Burnt to a toast**
This carbonized loaf of bread was one of several found in the brick oven of a bakery. The baker's stamp can still be seen, nearly 2,000 years after the day the bread was baked.

Modern Italian bread

Flour mill made of lava, a tough rock also used to pave streets

Portrait of a lady, detail of a floor mosaic found at Pompeii

Bowl of preserved eggs

## Snake charm
Fine gold and silver jewellery, some set with emeralds, was found in the buried town. This hollow bracelet in the shape of a coiled snake is made of thick gold.

Fresh walnuts

Fresh figs, still grown on the slopes of Vesuvius

Bowl of carbonized figs

# Charcoal

Objects that contain carbon, like wood or food, usually burn when heated. But when there is not enough oxygen for them to burn in the normal way, they turn to charcoal instead. This process, called carbonization, left many foodstuffs perfectly preserved in the ash.

Bowl of carbonized walnuts

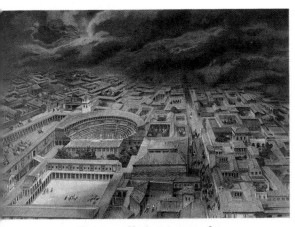

## Pompeii destroyed
The large theatre (the semi-circular building) and the gladiator's gymnasium (in front) can be seen in this artist's impression of the destruction of Pompeii.

## Death of Pliny the Elder
In a letter, Pliny the Younger wrote of his uncle and an official fleeing with "pillows tied upon their heads with napkins; and this was their whole defence against the storm of stones that fell around them... my uncle... instantly fell down dead; suffocated... by some gross and noxious vapour... his body... looking more like a man asleep than dead."

## The faithful dog

This guard dog died chained to his post, still wearing his bronze collar.

# Caught in the act of dying

More than 2,000 people died in Pompeii during the eruption. As the fleeing Pompeiians fell, the rain of ash and pumice set around their bodies like wet cement. Over time, the soft body parts decayed and the ash and pumice turned to rock. The shapes of the bodies were left as hollows in the rock with only the hard bones remaining. In 1860, the Italian king appointed Giuseppe Fiorelli as director of the excavations. Fiorelli invented a method for removing the skeletons from the hollows and filling the space with wet plaster of Paris. After the plaster hardened, a true representation of the bodies could be dug out of the rock. Many of these startling casts show people grimacing or huddling together in terror.

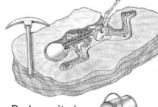

Body cavity is discovered

Cavity is filled with wet plaster of Paris

## Last day of Pompeii

This fanciful painting of the destruction of Pompeii by 19th-century German artist Karl Bruillov shows houses catching fire.

Cast of suffocated baby, found in the Garden of the Fugitives

## Shroud of death

This cast shows the folds of the dead man's clothing. He is clutching his chest, suggesting that it was painful to breathe. Most of the victims died by suffocating

*Part of woman's skull shows through cast*

Cast of man who died shielding his face with his hands

### Killed on duty

The American writer Mark Twain was most impressed by the remains of a soldier in Pompeii who had stayed at his post during the eruption.

### Mother and child

This mother was trying to shield her child when they were overcome. They were found with several other families in the Garden of the Fugitives.

### Health warning

This skeleton mosaic found near Pompeii is a "memento mori", a reminder of death.

Fiorelli takes detailed notes while supervising an excavation

*Pyroclastic flow deposit*
*Pyroclastic surge deposits*
*Ash and lapilli*

### Layers

The layers of ash, stones, and lava can still be seen.

# Herculaneum

When Vesuvius erupted in 79 CE, the ash cloud that engulfed Pompeii missed Herculaneum (p. 26). Less than 3 cm (1 in) of debris had fallen on the town when it was blasted by a great surge of hot ash and gas. Early excavations uncovered very few bodies, which was puzzling. But in the 1980s, several hundred skeletons were found huddled together beneath huge brick arches that once stood on the shoreline.

**Neptune and Amphitrite**
This mosaic of two Roman gods was unearthed at a merchant's house in Herculaneum.

**Walking in the ruins**
The excavations of the Roman town have created a deep hole that is surrounded by the modern city of Herculaneum (p. 60). This street is laid with lava paving stones.

### Roman skeletons
Unlike the bones in Pompeii, the skeletons from Herculaneum have no surrounding body shape. This is because they lay in waterlogged ground and the wet ash packed tightly around the bones.

### A tomb of hot rock
Herculaneum was hit by six pyroclastic surges (pp. 16–17) followed by thick flows of hot ash, pumice, and rock. These flows buried the town in 20 m (66 ft) of volcanic debris.

1631 eruption

### Text book eruption
This 1767 engraving (above), which probably shows the 1760 eruption, was published in an 18th-century text book.

### Hamilton's view
Lord Hamilton (p. 21), included this view of the 1779 eruption in his book *The Campi Phlegraei*. The artist is Pietro Fabris (p. 39).

# The most visited volcano

The Romans who lived in the shadow of Vesuvius were scarcely aware that it was a volcano. The mountain had erupted 800 years earlier, but it had been calm since. The biggest recent eruption, in 1631, produced pyroclastic surges and flows. In the 18th century, travellers flocked to Naples to visit the angry mountain. Even today tourists climb to the summit and pay to look into the steaming crater.

German etching of 1885 eruption showing fires started by lava flows

### On the tourist map
This cartoon shows English tourists at Vesuvius in 1890. A guidebook of 1883 advises sightseers to wear their worst clothes because boots could be ruined by the sharp lava and dresses stained by the sulphur.

Vesuv. Ash rain of the eruption (March 1944: days 22. 23. 24. 25. 26)

Photograph of tourists watching the 1933 eruption

### Souvenir of Vesuvius
Centuries ago, souvenirs from Naples included Roman artefacts stolen from the excavations. These days security is tighter, and boxes of lava and ash are more common souvenirs.

# St Pierre

One of the worst volcanic disasters of the 20th century happened on the Caribbean island of Martinique. On 8 May 1902, Mount Pelée – the volcano that towered over the city of St Pierre – erupted just before 8 am. The mountain sent a cloud of glowing gas down upon the port and all its inhabitants were engulfed. Within minutes, St Pierre was charred beyond recognition. A few sailors survived on their ships, but all but two of the city's 29,000 residents were killed.

**Broken statuette**

**Statue**
The heat pitted the surface of this statue. The other side (which faced the volcano) shows more damage.

**Alfred Lacroix**
French volcanologist Alfred Lacroix arrived in St Pierre on 23 June. In his report on Mount Pelée, he described strange "glowing clouds". Nowadays these are known as pyroclastic flows or surges (p. 16).

**Stopped clock**
This pocket watch was melted to a standstill at 8:15 am.

**Melted medicine bottle**

*Carbonized spaghetti*

*Carbonized prune*

*Ash fragment*

**Remains of mousetrap**

**Melted glass**
Discovered in the 1950s, these partly melted objects give a glimpse of everyday life in the small French colony at the beginning of the 20th century. Some objects are either so melted or so unfamiliar that it is hard to guess what they are.

*Fine volcanic ash melted into glaze*

**Melted wine bottle**

**Melted metal fork**
**(rust occurred after eruption)**

Top of charred human femur (thigh bone)

### Ruined city
The walls of some buildings were all that was left standing in St Pierre. Many died in the cathedral, where mass had just begun.

### Protecting angel?
This angel figurine, made of metal is just about recognizable.

Heap of glass melted beyond recognition

Charred mug

Squashed candlestick

### Petrified
Some objects containing carbon were scorched or burned completely. Others were carbonized (pp. 26–27).

*Carbonized coffee beans*

### Melted metal
Many metal objects melted in the heat. These iron nails fused together, while the metal spoon lost part of its bowl. The candlestick was probably squashed by a falling building. Copper telephone wires in the town were not melted, so the temperature of the cloud must have been less than 1,083°C (1,981°F), the melting point of copper.

Heap of fused iron nails

### Eternal figure
The wooden cross was burned off this crucifix, leaving just the figure of Jesus.

Fused coins

### Out of the frying pan
One of the two survivors in St Pierre was Auguste Ciparis, a condemned prisoner. He survived because his cell had thick walls and a tiny window. He was later pardoned and became part of a circus act.

Melted metal spoon

# Affecting the world's weather

A big ashy volcanic eruption has a dramatic effect on the weather. Dark days, severe winds, and heavy rain may plague the local area for months. If the gas and dust are thrown high into the atmosphere, they can spread round the world, affecting the climate of the whole planet. The volcanic material filters out some sunlight, reducing temperatures down below. In the longer term, volcanic particles may cause global cooling, mass extinctions, or even ice ages.

**Early Earth**
About 4,000 million years ago, planet Earth had no atmosphere and its surface was covered with erupting volcanoes. All the water in the oceans and many of the gases that make up the atmosphere have been produced by volcanic eruptions.

**Little Ice Age**
Two major eruptions in Iceland and Japan in 1783 led to several icy winters in Europe and America.

**Volcanic sunset**
Eruptions can cause unusually red sunsets. This sunset was caused by dust from the 1980 eruption of Mount St Helens (pp. 14–15).

**Dinosaurs**
It is likely that the dinosaurs died out due to climate change caused by a major asteroid impact and huge volcanic eruptions that occurred around 65 million years ago.

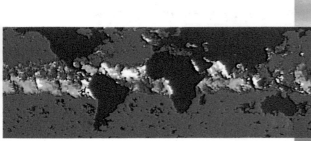

## Floating around the globe
The June 1991 eruptions of Mount Pinatubo in the Philippines (right and p. 17) spewed ash and gas into the stratosphere. Satellite images (above) showed that by 25 July the particles had spread around the world.

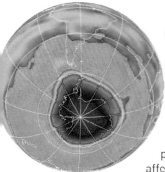

## Ozone hole
This false-coloured satellite image shows the hole in the ozone layer over the Antarctic. Sulphur particles thrown up by Pinatubo may cause further damage to this protective layer. This could affect world temperatures.

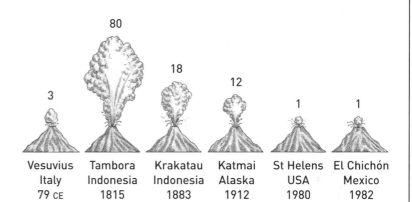

| 80 | 18 | 12 | | 1 | 1 |
|----|----|----|----|----|----|
| 3 | | | | | |

| Vesuvius Italy 79 CE | Tambora Indonesia 1815 | Krakatau Indonesia 1883 | Katmai Alaska 1912 | St Helens USA 1980 | El Chichón Mexico 1982 |

## Comparing the size of eruptions
The amount of ash a volcano spews out is a good measure of the size of the eruption. This diagram compares the amounts of ash produced in six major eruptions. The units are cubic km.

An artist's impression of the 1883 eruption of Krakatau, Indonesia

## Krakatau
In 1883, the Indonesian island of Krakatau was blown to pieces in a massive eruption (p. 57). The explosion was heard 4,000 km (2,485 miles) away in Australia. Floating islands of pumice drifted across the Indian Ocean for months afterwards. This piece was washed up on a beach in Madagascar, 7,000 km (4,350 miles) away.

# In hot water

In volcanic areas, heat from the rocks also heats the water in the ground. During long dormant (inactive) periods, the hot water may shoot to the surface in steam vents, geysers, hot springs, and pools of bubbling mud. These hydrothermal (hot water) features can be put to good use. Steam can be used to generate electricity, and hot groundwater can help to heat homes and greenhouses.

**Vulcan, god of fire**
The ancient Romans believed Solfatara volcano in Italy was one of the workshops of the blacksmith god Vulcan – hence our word "volcano".

**Measuring the Earth's heat**
An instrument called a thermocouple (p. 43) is being used to measure the heat of a steam vent, or fumarole, in the Solfatara crater. Temperatures here can reach 140°C (284°F). Changes in the heat and gas can give clues to future volcanic eruptions.

**Healing powers**
The fumaroles in Solfatara give off acid gases as well as steam. The steam that emerges from the caves is meant to have miraculous healing powers. Since Roman times, visitors have taken steam baths to treat arthritis and breathing problems.

**Crystals of sulphur**
The sulphur in volcanic gas cools and forms crystals. These huge, yellow crystals are from Sicily, where sulphur has been mined for centuries. Sulphur has many uses, particularly in manufacturing. It is added to rubber to make it more durable in a process called vulcanization – named after the Roman fire god Vulcan.

## Bubbling mud

Some fumaroles bubble up through a mud bath of their own making. The acid sulphur gases eat into the rock they pass through, creating a pot of soft mud. The mud in this pot at Solfatara is 60°C (140°F). Some mud pots are much hotter, while others are cool enough for people to relax and bathe in.

## Smelly gas

Sulphur crystals can be clearly seen around this fumarole vent. Close to the vent, the hot, smelly gases are invisible. Like the steam from a kettle, they show up only when the water vapour begins to condense (turn into water) a few centimetres away.

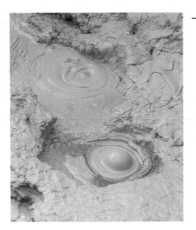

## Roman baths

The ancient Romans built huge public baths with hot running water. Baths fed by natural hot springs became medical centres where sick people came to bathe in the mineral-rich water.

Crust of tiny sulphur crystals
from fumarole in Java, Indonesia

Souvenir plate
showing Old Faithful
Geyser, USA (p. 7)

## Hot water power

About 40 per cent of Iceland's electricity comes from hydrothermal power stations. Countries such as Japan and the US are also developing hydrothermal power programmes.

OLD FAITHFUL GEYSER

# Sleeping beauties

Volcanoes sometimes sleep (lie dormant) for years, or even centuries, between eruptions. In this dormant period, volcanic gases may seep gently from the magma beneath the volcano. As these gases rise through the rocks of the volcano mountain, they react chemically with minerals already in the rocks to create new minerals. These are often brightly coloured with large crystals. At Earth's surface, the gases fume gently off into the atmosphere.

Church built on eroded remains of old volcano, Le Puy, France

Radiating zeolite crystals from the Faëroe Islands

## Born in the lava

Zeolite crystals grow in old gas bubbles in lava. They are found in a variety of colours and forms.

## Agates

These beautiful banded stones form in cavities in cooled or cooling volcanic rocks. Each band is formed at a different time period.

*Outer layers of this agate are oldest*

Adventurers descend into the crater of Hekla, Iceland, in 1868

## Crater lake

Craters often fill with rainwater. This crater lake, on the Shirane volcano in Japan, is very acidic due to gas seeping up from the magma chamber. In an eruption, the acidic water mixed with hot rock and debris could cause a deadly mud flow (pp. 56–57).

Brightly coloured rocks seen by Lord Hamilton at Solfatara (pp. 36–37) and illustrated by Pietro Fabris

Olivine from St John's Island in the Red Sea

Cut diamond

# Precious stones

Chemicals in the hot volcanic fluids cool slowly inside gas bubbles or in other cavities in the volcanic rock. This process often produces large crystals that can be cut and polished into gemstones. Hard stones, like diamonds, are the most prized because they last forever.

Uncut diamond in volcanic rock from the mantle, from Kimberley in South Africa

Cut peridot

## Red Sea gem

Gem-quality olivine is a deep green colour. The gem is known as peridot.

Agate-lined geode (cavity) from Brazil

## Birth of a caldera

During a large ashy eruption, the empty magma chamber may not be able to support the volcano's slopes. These collapse inwards, leaving a huge dip called a caldera.

## Santorini

This Greek island is the rim of a caldera formed by a huge volcanic eruption c. 1620 BCE. The massive explosion may have led to the collapse of the Minoan civilization on the neighbouring island of Crete.

## Sleeping volcano

Mount Rainier is one of a chain of volcanoes in the Cascade range, USA. There are no records, but the volcano probably erupted several times in the 1800s. These events can be dated from tree rings, which show stunted growth after an eruption.

# Life returns

Volcanic eruptions can have a dramatic effect on the landscape. Volcanic ash is full of nutrients that enrich the soil, but too much can be catastrophic for farmers. Thick, sticky lava covering the land can take months to cool. Decades may pass before plant life returns to the landscape. Only when a rich soil covers the ground is it lush and fertile again. This process may take generations.

1944 lava flow, Mount Vesuvius

Raw lava

*Dense, interlocking crystal structure*

A few lichens find a home on the lava

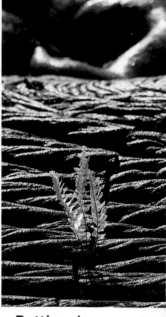

## Putting down roots

Ferns, mosses, and lichens are some of the first plants to grow after an eruption. Here, a fern takes root in a ropy pahoehoe lava flow less than a year old on the slopes of Kilauea, Hawaii (pp. 22–23).

*Grasses, often the first flowering plants to grow*

*Beginnings of topsoil*

*Lichen cling to exposed parts of rock*

Lichen covers the lava, providing a soft surface for other plants

### Gathering moss

The rate at which plants grow back after an eruption depends on the type of erupted material. Plants are slowest in taking root on lava flows; they grow more quickly in ashy, pyroclastic material (p. 14). These lava pieces are all from the 1944 aa flow on the west slope of Mount Vesuvius in Italy. Some 47 years later, lichen covers a lot of the flow, and moss, grasses, and weedy flowering plants are taking root.

Rock breaks down to soil, and grass and moss take root

*Two species of moss grow in thin soil*

**Mount Vesuvius steaming after mild eruption of 1855**

*New cone is still bare ash*

*Monte Somma, part of caldera (p. 39) left by huge, prehistoric eruption*

*Pine forests cover lower slopes*

### Birth of an island
In November 1963, an undersea eruption off southwest Iceland gave birth to a new island, Surtsey (p. 20).

### Washed ashore
Seeds blown over or washed up on the beach of Surtsey soon took root in nearby ashfields (above).

### Lacrima Christi
Mount Vesuvius is shown on the label of this wine grown on the volcano's slopes. Without the nutrients from the volcanic ash, the vines would not grow so thickly and the wine would taste less sweet.

Peacock butterfly lives on nectar of flowering plants

*Weedy flowering plant*

Eventually, the soil is thick enough to support larger plants

### Flower of Lydia
This brilliantly coloured shrub, a kind of broom, is one of the first plants to grow on the lava at Vesuvius.

### Through the grapevine
The land around Vesuvius has been fertilized by ash from regular eruptions over the last 20 centuries. Grapes grown in the lush soil are used to make wine.

### Roman amphorae
The stacks of amphorae for storing wine and olive oil found at Pompeii (pp. 26–31) show how fertile the soil was in Roman times.

Mosaic of Venus, Roman goddess of fertility, found in Pompeii

# Studying volcanoes

Volcanologists – scientists who watch, record, and interpret volcanoes – spend years monitoring volcanoes to try and predict when and how they will next erupt. Most of their work involves analysing data, but fieldwork on the slopes of active volcanoes is vital. This involves taking lava and gas samples and measuring changes in landforms and temperature – usually while wearing protective clothing.

### Katia Krafft
French volcanologists Maurice and Katia Krafft (left) spent their lives documenting volcanoes. Sadly, the husband and wife team were killed during the eruption of Mount Unzen in Japan in 1991.

### Spaced-out suit
This protective suit has a metal coating that reflects the heat of the volcano and leaves the person inside cool. Heat-proof boots are worn to walk across the red-hot lava.

### Volcano notebook
The volcanologist makes notes and sketches of everything that takes place during an eruption. The significance of some things may only become clear later.

### Hot rod

This metal rod is ideal for collecting red-hot lava. The volcanologist dips the end into the lava flow, then twists it around, hooking up a blob of lava.

*Hard hat*

### Tape measure

A tape measure is handy to check cracks in the ground that may widen by a minuscule amount each day.

Binoculars

*Gloves made from the heat-resistant mineral asbestos*

### Too hot to handle

Volcanologists wear asbestos gloves to pick up red-hot lava. Hard hats protect against volcanic bombs (p. 18).

### Pathfinder

This mining transit (left) is a useful tool for mapping the ground of a volcano. It has a compass and a spirit level (to find verticals and horizontals). Small and light, it can be clipped onto the volcanologist's belt.

### A closer look

Binoculars allow people to observe a volcano from a safe distance.

*Spirit level*

*Compass*

*Rotating stage*

### Mapping the moving Earth

A precise spirit level is used to detect the small changes in ground level that occur before an eruption.

*Folding, portable tripod*

Thermometer reading up to 250°C (482°F)

### Taking the volcano's temperature

Katia Krafft takes the temperature of a lava flow on Piton de la Fournaise volcano, Réunion (pp. 22–23). She is using an electric thermometer called a thermocouple. The reading was 1,100°C (2,012°F).

# Other planets

The many space missions over the past few decades have brought back photographs of volcanic activity on other planets. Like Earth, the Moon, Venus, and Mars have surfaces that have been partly shaped by volcanic activity. The volcanoes on the Moon and Mars have been extinct for many millions of years. Scientists suspect that Venus's volcanoes may still be active. But of all the other planets in our solar system, only Io, one of Jupiter's 16 moons, shows volcanoes that are still erupting.

### Tidying the planet
The hero of Antoine de Saint-Exupéry's children's story *The Little Prince* lives on a planet with two active volcanoes. Before setting out on a journey, he cleans them out to be sure they won't erupt and make trouble while he's away.

*Molten rock*

*Lava flow*

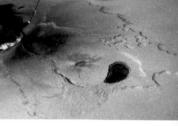

### Surface of Io
This coloured infrared image shows a volcanic region on the surface of Io. The image shows molten rock and a 60-km- (37-mile-) long lava flow, produced by a volcanic eruption.

### Olympus Mons
The extinct volcano Olympus Mons is 600 km (373 miles) across and rises 25 km (16 miles) above the surrounding plain. It is the largest volcano to be found in the universe so far. Huge calderas (p. 39) nest inside one other at its summit.

*Clouds of ice shroud the summit*

*Volcano Gula*

*Volcano Sif*

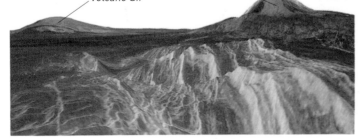

### Erupting into space
This image shows the volcano Prometheus erupting on Io. Clouds of gas rose 160 km (100 miles) above the surface. The eruption clouds shoot far into space because Io has very low gravity and virtually no atmosphere.

### Beneath the clouds
The spacecraft *Magellan* used imaging radar to penetrate the dense atmosphere of Venus. The images revealed huge volcanoes lurking beneath the clouds. They were named after women, including goddesses from mythology.

*Vidicon camera viewer*

## Space voyagers
The two *Voyager* spacecrafts were launched in 1977. They flew past Jupiter in 1979 and Saturn in 1980–1981.

*Model of
Voyager 1*

*Lava flows*

*300-km-
(186-mile-)
high gas
plume from
volcano
Pele*

*Dark,
inactive
volcano,
Babbar
Patera*

## Shooting the stars
The *Voyager* crafts caught eight of Io's volcanoes in the act of erupting. They also saw about 200 huge calderas, some filled with what seem to be active lava lakes. The images were collected by Vidicon, a type of TV camera.

*Propulsion
fuel tank for
making delicate
adjustments to
flight path*

# When the earth moves

1906 cartoon, captioned "I hope I never have one of those splitting headaches again."

Being in a large earthquake is a terrifying experience. When the shaking starts there is no knowing how long it will go on or how severe it will be. The longest tremor ever recorded, the Alaskan earthquake of 23 March 1964 lasted four minutes. But most quakes last less than a minute. In those brief moments, homes, shops, and even entire cities are destroyed. Afterwards, great cracks may appear in the ground. Aftershocks, which follow a big tremor, can continue for months.

### Disaster movie
This earthquake movie was shown in "Sensurround" – low frequency sounds meant to simulate earthquake shaking.

### Shaken to the foundations
Almost 75 years old, these wood buildings in San Francisco slipped off their foundations during the 1989 earthquake (p. 7).

### Panic sets in
People leave buildings and rush into the streets in panic as an earthquake shakes the city of Valparaiso, Chile, in 1906.

### Folded
This book was damaged in an earthquake that devastated Skopje in Yugoslavia in 1963. Skopje sits on the site of the ancient city of Scupi, which was completely flattened by an earthquake in 518 CE.

*Temple of Jupiter*

## Rocking the temple

The Roman town of Pompeii was hit by a large earthquake in 62 CE. A marble carving from a house in the town shows the damaged Temple of Jupiter.

## When the earth breaks

Solid rocks can fracture when the earth shakes. This road cracked during an earthquake measuring 6.9 on the Richter scale (pp. 48–49).

## Cracking ground

Volcanic tremors are caused by moving magma. Here, rising magma has cracked the ground before an eruption of Piton de la Fournaise volcano, Réunion.

## Shaken up

The Roman philosopher Seneca wrote about the earthquake that damaged Pompeii in 62 CE. He was particularly interested in the psychological effects of the ground shaking and the natural fear it caused.

## Voltaire

The French writer Voltaire (1694–1778) wrote about the huge Lisbon earthquake of 1755 in his novel *Candide*. The novel made fun of religious figures who said that God was punishing the residents for their immoral ways.

## Solid as a rock?

This piece of limestone has a natural polish caused by earthquake stresses and strains. The flat surface was almost melted by the heat generated as the rock broke.

## Shaking, fire, and flood

The 1755 quake destroyed three-quarters of Lisbon's buildings. Huge tsunamis (pp. 56–57) destroyed the harbour, and more than 10,000 people were killed.

# Intensity and magnitude

How do you measure the size of an earthquake? News reports usually give the quake a magnitude on the Richter scale. The Richter magnitude is useful because it can be worked out from a recording – called a seismogram – of the earthquake waves (pp. 52–55). These waves can be recorded from anywhere on the planet. The intensity of the shaking and how it affects buildings and people cannot be recorded by a machine. It is compiled by inspecting damage and measured on a scale such as the Modified Mercalli Intensity Scale.

Giuseppe Mercalli
(1850–1914)

## Intensity

The Italian volcanologist Giuseppe Mercalli created his intensity scale in 1902. He used 12 grades with Roman numerals.

**I** The shaking is not felt by people, but instruments record it.

**II** People at rest notice the shaking (above), especially if they are on upper floors of buildings. Lightly suspended objects may swing.

**III** People indoors feel a vibration like the passing of a light truck. Hanging objects start to swing (above).

**IV** Vibration like a heavy truck passing. Dishes rattle and standing cars rock.

**V** Felt outdoors. Liquid in glasses slops out (above). Small objects fall.

**VI** Felt by all. Many are frightened and rush outdoors. People walk unsteadily; dishes and windows break (above).

# Magnitude

In the 1930s, Charles Richter wanted to compare the sizes of local earthquakes. First, he calculated the distance between himself and the earthquake. He then used this distance, together with the wiggly tracings of the ground movement (pp. 52–55) to come up with the magnitude.

American seismologist Charles F. Richter (1900–1985)

Intensity contours for an earthquake that struck Japan on 22 May 1925

*Epicentre*

III IV V VI II

## Recording shakes

Richter took the smallest earthquake he could record at the time and called it magnitude zero. The highest Richter magnitudes recorded are about 9.

**II** Difficult to stand (above). Plaster and loose bricks crack and fall. Waves on ponds. Bells ring.

**VIII** Steering of cars affected. Damage to walls, some of which fall. Falling chimneys, steeples (above). Cracks in wet ground.

**IX** Panic. Animals run in confusion. Frame buildings, if not bolted down, shift off their foundations (above). Mud and water bubble out of ground.

**X** Masonry and frame buildings destroyed (above). Some well-built wooden buildings destroyed. Large landslides. Water thrown out of rivers.

**XI** Railway lines greatly distorted. Underground pipelines out of service. Highways useless. Large cracks in ground. Large rockfalls.

**XII** Practically all built structures destroyed or useless (above). Ground much altered with cracks. River courses moved. Waterfalls appear.

# Shock waves

Earthquake waves travel fast – about 25,000 kph (15,534 mph) in rock. In the seconds after the rock fracture that causes the earthquake, shock waves travel out in all directions. Usually they are most devastating near the epicentre – the place on the Earth's surface nearest to where the rocks have fractured. But sometimes the waves are slowed down by soft sands and muds. This can cause severe shaking even far from the epicentre.

## Seismogram

This is a recording of a 5.1 magnitude quake. The primary (P) waves arrive first, followed by slower secondary (S) waves. The time lag between the P and S waves – 17 seconds – is used to calculate the distance from the epicentre.

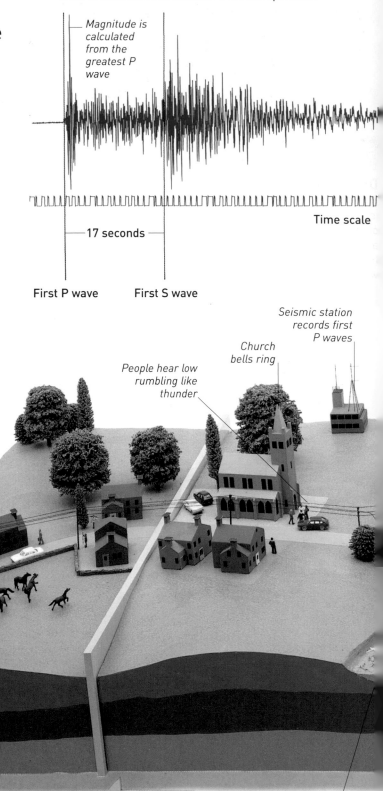

*Magnitude is calculated from the greatest P wave*

Time scale

— 17 seconds —

First P wave   First S wave

Seismic station records first P waves

Church bells ring

People hear low rumbling like thunder

Animals are restless, and may rush about and cry

## Epicentre

In 1989, seismologists (people who study earthquakes) in Scotland, Africa, and India calculated how far away a quake had struck and drew a circle across the globe based on their results. The circles met in the Caspian Sea, the epicentre of the quake.

Eskdalemuir, Scotland

Epicentre in Caspian Sea

Hyderabad, India

Lusaka, Zambia

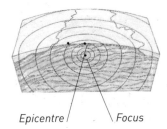

*Epicentre*   *Focus*

## Deep focus

An earthquake's focus – the area where the rocks have fractured – is usually many kilometres inside the Earth.

### Before the earthquake

Animals often sense something is wrong in the minutes before an earthquake strikes.

P waves

Startled by ripples, waterbirds fly off ponds

### The first waves strike

The first P waves to arrive may be s small that they are heard but not fel

Woman sits by the ruins of her house in Lice, Turkey

## Devastated in a minute

In 1843, the town of Pointe-à-Pitre on the island of Guadeloupe was hit by an earthquake of magnitude 8. The shaking lasted for about a minute – long enough to reduce most of the buildings to ruins. A fire that followed destroyed what was left of the town.

# Living through an earthquake

This model shows the waves from a large earthquake. The fast P waves (yellow) are about to strike the area on the far left. S waves (blue) follow, causing considerable damage. The slowest, surface waves (red), arrive seconds later, causing the total collapse of buildings already weakened by the S waves.

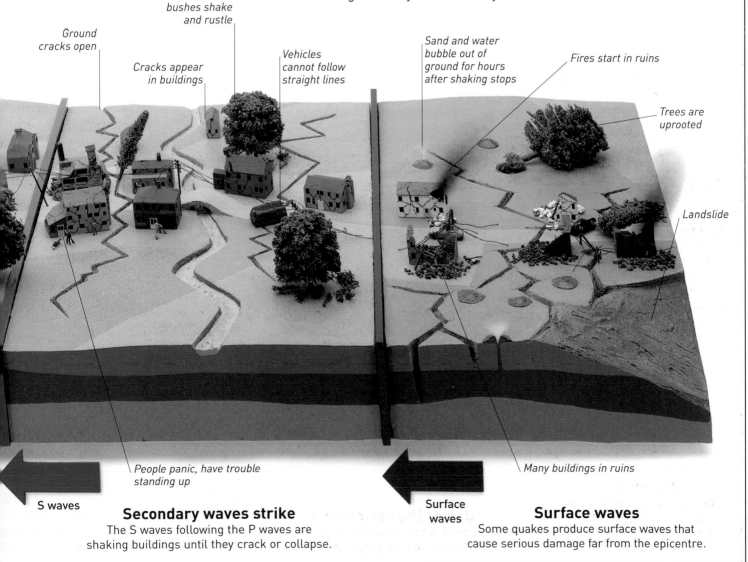

Ground cracks open

Trees and bushes shake and rustle

Cracks appear in buildings

Vehicles cannot follow straight lines

Sand and water bubble out of ground for hours after shaking stops

Fires start in ruins

Trees are uprooted

Landslide

People panic, have trouble standing up

Many buildings in ruins

S waves

Surface waves

### Secondary waves strike
The S waves following the P waves are shaking buildings until they crack or collapse.

### Surface waves
Some quakes produce surface waves that cause serious damage far from the epicentre.

# Measuring earthquakes

The first instrument for recording earthquakes was built by the Chinese scientist Zhang Heng in the second century CE. The device, known as a seismoscope, could record earthquakes too slight to be noticed otherwise. In 1856, a more sophisticated device was invented by the Italian Luigi Palmieri. His seismograph was designed to measure the overall size of the earthquake shaking (pp. 48–49).

### Early seismologist

The Chinese were keeping lists of earthquakes as early as 780 BCE. But it was not until 132 CE that the Chinese astronomer Zhang Heng (78–139) invented the first seismoscope. The bronze device measured about 2 m (6.6 ft) across.

Inner workings of Zhang Heng's seismoscope

Suspension mechanism pulls on dragon's mouth

Pendulum

### Seismoscope

During a tremor, the vessel moves more than the heavy pendulum hanging inside. This triggers one or more of the dragons to open their jaws and release a bronze ball into the mouths of the toads below.

Ball held in dragon's mouth

The toad that is farthest from the epicentre catches the falling ball. This indicates which direction the quake came from

## Ring my bell

This reconstruction of a seismoscope (left) was built by Italian naturalist and clockmaker Ascanio Filomarino in 1795. When the ground shook, the pendulum stayed still. But the rest of the apparatus shook, making bells ring and a clock tick.

*Clock starts ticking when shaking begins. If it is found at 3 pm, the quake began three hours earlier*

*Heavy weight of pendulum*

*Bells ring when quake starts*

*Pencil leaves trace on paper*

*Clock stops when shaking starts*

*Ticker tape*

**Recording apparatus of Palmieri's seismograph, which produces a ticker tape record of the shaking**

*Four mercury-filled U-tubes make contact with platinum wires held just above the liquid metal*

*Ivory pulleys which only move in one direction record the maximum size of the movements*

*Wire of pendulum that records direction of tremor*

*Spring bounces in vertical movement of earthquake*

*Electrical circuit is completed when platinum point dips in the dish of mercury below*

## Palmieri's seismograph

The first seismograph was built in 1856 by Luigi Palmieri (1807–96). The larger part (right), which contains tubes of mercury, detects earthquakes. The second part (top right) prints a record of them.

### Observatory

Palmieri developed his device while he was director of this observatory near the crater of Vesuvius.

*Weight of pendulum*

### Luigi Palmieri

Palmieri realized that an instrument that measured ground tremors might help to predict eruptions. This instrument was the result of his experiments with electricity.

*Maximum wave size recorded here*

**Tokyo, 1923**
This photo shows the city of Tokyo after the huge quake of 1923. At least 200,000 died in the fire storm that followed.

Seismogram of 1923 Tokyo earthquake, recorded by the Gray-Milne seismograph in Oxford, England

**John Milne**
An English geologist, John Milne (1850–1913), invented his own seismograph while he was teaching geology in Tokyo.

# Seismometers

Seismometers capture earthquake movement. They work on the principle that an earthquake shakes a heavy pendulum less than the surrounding ground.

**Side view of Gray-Milne seismograph**

*Three pens record vertical movement and two types of horizontal shaking*

*Pendulum to record horizontal shaking*

**Ground motion**
This seismograph, designed by Thomas Gray and John Milne, had three pendulums and three pens, to record the three types of ground motion – vertical, east-west horizontal, and north-south horizontal.

*Clock shows the moment the quake starts*

## A modern observatory

At the Vesuvius Observatory, great reels of paper record the ground movement measured by a series of seismometers. Their data is beamed into the recording centre by radio or along telephone lines. Modern seismographs record on magnetic tape, which allows for much better analysis.

## Portable

Networks of portable seismometers are used to monitor the aftershocks of big quakes and ground tremors during volcanic eruptions.

## Moonquakes

American astronauts left seismometers on the Moon to record moonquakes. Many moonquakes are caused by meteorites hitting the surface.

*Paper drum winds very slowly between earthquakes. When shaking starts, gears change and the drum starts feeding the paper through much faster*

*Case hides suspended pendulum*

*Handle for winding up weight that turns drum*

*Damping system, which makes sure that each shock wave is only recorded once*

*Suspended weight drives paper drum (mechanical clocks driven in same way)*

*Smoked paper seismogram*

*Arm from which pendulums are suspended*

## Smoking up

Early seismographs, many still in use, scratch their traces on smoked paper. This avoids the problem of ink running out – a disaster during tremors.

## Heavy duty

This is a restored version of the seismograph invented in 1908 by German scientist Emil Wiechert (1861–1928). Its 200 kg (440 lb) mass measures the two horizontal movements of ground shaking. It worked with a smaller instrument that measured vertical motion.

# Mud, fire, and floods

The events that follow an earthquake or volcanic eruption can be even more dangerous than the disaster itself. Heavy rain mixed with volcanic ash can create devastating mud flows. In mountains, both quakes and eruptions may trigger landslides and avalanches; by or beneath the sea, they can both cause giant water waves known as tsunamis.

**Swept away**
The ashy eruptions of Mount Pinatubo in the Philippines (p.17) were accompanied by mud flows that swept away roads, bridges, and several villages.

Overview of Armero mud flow, 1985

**Buried in mud**
In 1985, the Ruiz volcano in Colombia, South America, spewed clouds of ash and pumice onto snow and ice at the mountain's summit. The melted snow and ash formed a heavy mud flow that travelled at speeds of up to 35 kph (22 mph). In the city of Armero, 60 km (37 miles) away, some 22,000 people were buried alive by the waves of mud, rock, and debris that set around them like wet concrete.

Lorry trapped in mud, Armero

### Pozzuoli

The town of Pozzuoli near Naples, Italy, has been shaken by many small earthquakes. Part of the town was abandoned after shaking damage in 1983. The town has risen several metres since then, so the harbour had to be rebuilt lower down.

*Old mooring post*

*New dock level*

### Dwarfing Fuji

Tsunamis are caused by both volcanic eruptions and earthquakes. The volcano Fujiyama in Japan is shown in this picture of a tsunami by Katsushika Hokusai (1760–1849).

### Tsunami

In 2011, an earthquake with 9.0 magnitude hit Oshika Peninsula in Japan, causing a tsunami. More than 18,000 people were killed. The tsunami also caused a system failure in the nuclear power plant Fukushima Daiichi. As a result, tonnes of radioactive water leaked into the Pacific Ocean.

### Krakatau, west of Java

Tsunamis as high as 30 m (100 ft) crashed into surrounding islands after the eruption of Krakatoa (now called Krakatau, p. 35). More than 36,000 people were killed.

### Avalanche

In 1970, a magnitude 7.7 quake off the coast of Peru caused a disastrous slide of snow and rock. The avalanche fell 4,000 m (13,123 ft) and killed more than 50,000 people in the valley below.

### Fire in the ruins

These firefighters are putting out a blaze after the 1989 San Francisco earthquake (p. 7). The fires that follow quakes or eruptions can completely destroy cities. If gas mains are broken or inflammable liquids spilt, the slightest spark causes fire. Shaking often damages water supplies, making a blaze harder to fight.

# State of emergency

The chaos that follows a big earthquake or volcanic eruption makes rescue difficult and dangerous. Half-collapsed buildings may topple further at any moment. Hazardous substances could suddenly catch fire or explode. In ash-flow or mud-flow eruptions, no one knows when to expect another surge. Damage to electricity, gas, and water supplies and disruption to communication links make rescue operations even harder to mount.

**Perilous rescue**
A survivor is lifted by helicopter in Armero, Colombia, in 1985 (p. 56).

**Muddy escape**
A survivor is rescued from the boiling mud flows that engulfed Armero in 1985. Many survivors had to be treated for burns.

*Controls showing level of infrared radiation*

*Strap worn around neck ensures expensive camera is not dropped in rubble*

**Finding live bodies**
A thermal image camera is used to locate people trapped after an earthquake. Survivors are often buried, wounded or unconscious, in the rubble of their collapsed homes. The camera uses infrared radiation to detect the heat of a living person.

**Helping out**
The London Fire and Civil Defence Authority sends trained teams to disaster zones like northwest Iran after the massive quake of June 1990.

*Searcher wears headphones to listen for human sounds in the wreckage*

## Trapped person detector

This device was used to find people trapped in wreckage after the Armenian earthquake of 1988. The device works by detecting vibrations.

Italian newspaper illustration from 1906 showing a boy being rescued from the rubble

GAIN
50
40
30
20
10
0
ON
PHONES
GAIN
LISTENING
L
L+R
R
TRAPPED PERSON LOCATOR
TPL 310 B
FILTER
LOW
MED.
HIGH
FREQUENCY
VOLUME
SENSORS
L
R
10
9
8
7
6
5
4
3
2
1
12 V
PEAK
HOLD
RST
CAROL

*Microphone, so rescuer can talk to trapped person*

*Red two-way electrode allows rescuer to talk to survivor*

*Yellow one-way electrode picks up vibrations*

### Haiti earthquake

Thousands of people were killed in Haiti in January 2010, when a magnitude 7 earthquake hit the country. Many noted buildings, such as the Presidential Palace and the National Assembly Building, were badly damaged.

### Sensitive nose

Sniffer dogs can also help to find survivors after an earthquake. Rescuers are often at risk as they work in ruined buildings. If an aftershock causes further collapse, the rescuers may have to be rescued, too.

# One step ahead

As the planet's population increases, more people are living in danger zones, along faults or close to active volcanoes. We cannot hope to stop disasters entirely, but we can reduce their number and scale. Learning to live in disaster zones means monitoring volcanoes and fault lines and building cities that can withstand earthquakes. It also means educating people to know what to do in an emergency.

Italian magazines produced for the 1990s, the International Decade for Natural Disaster Reduction

## Earthquake-proof cities

Many modern buildings in earthquake-prone cities are designed to withstand shaking. The Transamerica Pyramid in San Francisco, USA, looks precarious, but in a major quake, the structures at the base will reduce sway by a third.

## In the shadow of Vesuvius

Two thousand years after the volcano's greatest eruption (pp. 30–31), modern Herculaneum (above) is a thriving town.

## Frank Lloyd Wright

This American architect was a pioneer in the design of earthquake-resistant buildings. His Imperial Hotel in Tokyo survived the 1923 quake almost unscathed.

## Shake till they drop

Built in 1923, this Japanese shaking table was used to test models of buildings to see how they stood up to severe shaking.

### Most measured place

The town of Parkfield in central California lies above the San Andreas fault system. Seismologists (pp. 48–49) have predicted a major earthquake here. A laser measuring system is being used to detect movements along the fault. It can detect ground movement of less than a millimetre over a distance of 6 km (3.7 miles).

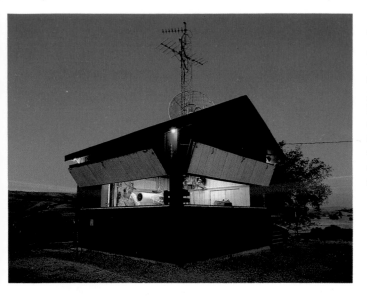

### Measuring creep

A technician for the US Geological Survey has been measuring creep – slow movement along the fault.

### Learning from the past

Earthquakes of the same size tend to happen in the same place at regular intervals. Studying large quakes – in this case, the one that rocked Pompeii in 62 CE (p. 47) – may help scientists to predict the next big tremor.

### Falling masonry

In this earthquake drill, rescue workers are treating actors "hit" by falling masonry. Designing buildings without heavy stone ornaments or chimneys might help to reduce the numbers of casualties like these.

Earthquake rescue practice in Japan

### Earthquake drill

In Japan and California, earthquake drills are part of everyday life. Children learn to keep a torch and shoes by their beds, so they can get to safety at night. The safest place indoors is under a piece of furniture like a table, or beneath a door frame.

# Anger of the gods

As long as people have lived on Earth, they have been curious about natural events like volcanic eruptions and earthquakes. Myths and legends are a way of recording or explaining these strange happenings. For centuries, many societies have explained natural events as the workings of a god or gods. It was thought that angry gods would punish people with a fiery eruption or the horrible shaking of an earthquake. Some societies still believe that certain gods live on the eerie summits of volcanoes, which are often shrouded in fire and cloud.

Christians in Naples, Italy, try to stop the 1906 eruption of Mount Vesuvius (p. 31) with crosses and prayers

### Human sacrifice
In Nicaragua, people used to throw young women into the lava lake at Masaya to stop the volcano from erupting.

Lava lake

### Popocatépetl
This Aztec illustration shows Popocatépetl in Mexico. When the volcano erupted violently in the 1520s, the Aztecs believed it was because the gods were angry with the Spanish conquerors who had looted their temples.

### Responsible frog
Many cultures believed that the ground they stood on was held up by some huge creature. The Mongolians believed this was a gigantic frog. Each time the animal stumbled under his great burden, the ground shook with an earthquake.

### One-eyed giant
From above, the craters on Mount Vesuvius look like giant eyes. They may have inspired the Greek myth of the Cyclops, a tribe of one-eyed giants who helped the fire god Hephaistos (below).

### Shaking the sea floor
When the Greek sea god Poseidon was angry, he banged the sea floor with his trident (spear). The ancient Greeks believed that this led to earthquakes and tsunamis (pp. 56–57).

Destruction of Sodom and Gomorrah, by an unknown Flemish painter

Bronze figure of Hephaistos, 1st or 2nd century BCE

### Sodom and Gomorrah
According to the Bible, God destroyed these cities with flood and fire because the inhabitants were evil.

### Master of fire
The ancient Greeks believed the god Hephaistos had his fiery workshops under volcanoes (p. 36). Another god, Prometheus, stole some of the fire from the volcanoes and gave it to humans.

### When the gods are away...
A Japanese myth says earthquakes were caused by a giant catfish. Normally the fish was pinned down by a large rock. But when the gods were away, the creature could escape. This print shows the gods flying back after a big quake, carrying a rock.

### Home of the gods
Mount Fuji in Japan is thought to be the home of the god Kunitokotache. The sacred spirit of the mountain, Fujiyama, is said to protect the Japanese people.

# Did you know?

## FASCINATING FACTS

Much of New Zealand's North Island, and all island life, were devastated by the eruption of the Lake Taupo caldera in 180 CE. Luckily, the first human inhabitants did not occupy the island until 1000 CE.

Around 200 black bears were killed in the eruption of Mount St Helens.

Figures from Mount St Helens estimate that 11,000 hares, 6,000 deer, 5,200 elk, 1,400 coyotes, 300 bobcats, and 15 mountain lions were killed by the blast.

Pyroclastic flows can travel at up to 500 kph (311 mph) and reach temperatures of 800°C (1,472°F), burning everything in their path.

A truck flees from the burning clouds of Pinatubo, Philippines, 1991

Where cracks form in the ocean floor, the water heated by magma can reach temperatures of 662°C (1,224 °F).

One of Iceland's greatest attractions used to be the Great Geyser, near Reykjavik, which had a jet 60–80 m (197–262 ft) high. The geyser stopped spouting in 1916.

When Krakatau in Indonesia erupted in 1883, the noise was so loud it burst the eardrums of sailors over 40 km (25 miles) away. Around 36,000 people died, most killed by tsunamis – some 30 m (100 ft) tall – that devastated Java and Sumatra. Villages, ships, and boats were swept inland.

Fires are a major problem after an earthquake. Fractured gas pipes mean fires spread rapidly and burst water mains dry up the hoses. This, together with streets blocked with debris, make the firefighters' work near impossible.

The greatest volcanic eruption in modern times was Tambora, Indonesia, in 1815. It produced 80 $km^3$ (19.2 $miles^3$) of volcanic ash, compared with 1 $km^3$ (0.24 $miles^3$) measured at Mount St Helens. In the past 10,000 years, only four eruptions have been as violent as Tambora.

A tsunami can travel at speeds of up to 600 kph (373 mph).

A steamer swept inland at Krakatau

Animals often act strangely before an earthquake. In 1975, in China, scientists correctly predicted an earthquake when they noticed snakes waking up from hibernation and rats swarming.

*Pyroclastic flows travelling at over 70 kph (40 mph) were recorded*

**Q** When is a volcano said to be extinct rather than dormant?

**A** A volcano is classified as active if it has erupted within the last few hundred years. It is dormant if it has not erupted in the last few hundred years but has erupted during the last several thousand years. If a volcano has not erupted during the last several thousand years, it is said to be extinct.

**Q** Have people ever tried to stop an advancing lava flow?

**A** When the volcano on the Icelandic island of Heimaey erupted in 1973, the residents tried to save the harbour on which the island depended from the lava. They did not stop it, but by spraying 6 million tonnes of sea water at the lava, they slowed it down and slightly altered its course. The lava flow stopped just 137 m (450 ft) from the harbour.

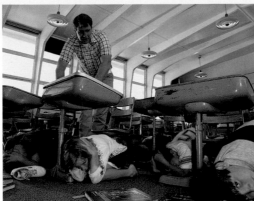

Earthquake drill at a school in the USA

**Q** What should you do if there is an earthquake?

**A** Shelter in a doorway or under a strong table and protect your head with your arms. When the tremors stop, leave the building, and shelter away from walls, which may collapse.

**Q** How many active volcanoes are there in the world?

**A** There are more than 1,500 active volcanoes that rise above sea level. On average, around 20–30 are actually erupting each month. Some of these are volcanoes that erupt continually, like Mauna Loa and Kilauea.

Mauna Loa, Hawaii

**Q** Could lava just come out of a crack in the earth?

**A** Yes, in 1943, in Paricutin in Mexico, a farmer found lava pouring from a crack in his field. Within a day there was a cone 10 m (33 ft) high. After a year of eruptions, the lava was 450 m (1,476 ft) tall and had engulfed the nearby town.

*A church tower rises from the lava of Paricutin*

Paricutin, Mexico, buried by lava

## Record Breakers

**BIGGEST VOLCANO**
Mauna Kea in Hawaii is a massive volcano, 1,355 m (4,446 ft) taller than Mount Everest, but much of it is under the sea.

**BIGGEST EARTHQUAKE**
In 1960, an earthquake of 9.5 on the Richter scale was recorded in Chile. It caused tsunamis that reached Japan.

**MOST DEVASTATING EARTHQUAKE**
In 1556, an earthquake of magnitude 8.3 killed 800,000 people in Shansi, China.

**HIGHEST TSUNAMI WAVE**
The highest tsunami wave on record struck Ishigaki in Japan in 1971. The monstrous wave was 85 m (279 ft) high.

**LARGEST FAULT SLIP**
The largest ever fault slip was the main cause behind the 2011 tsunami in Japan. The fault in the floor of the North Pacific Ocean slipped by 50 m (164 ft).

# Timeline

The timeline below includes just some of the major volcanic eruptions and earthquakes of the past four thousand years. We have most information about events in living memory or from records in recent history, but volcanologists can look much further back in time by studying the features of the Earth's surface.

Gas, dust, and rock explode from Mount St Helens, 1980

### c.1620 BCE Santorini, Greece

Violent eruptions buried the island of Santorini under 30 m (98 ft) of pumice.

### 79 CE Vesuvius, Italy

A burning cloud of volcanic ash engulfed the Roman towns of Pompeii and Herculaneum killing thousands.

### 1755 Lisbon, Portugal

A earthquake measuring 8.5 on the Richter scale reduced the city to rubble.

### 1783 Skaftar fires, Iceland

A fissure (crack) 27 km (16 miles) long spewed out lava and poisonous gases.

Craters mark Skaftar fissure today

### 1815 Tambora, Indonesia

The biggest eruption ever recorded. Around 90,000 people were killed. Volcanic dust reduced levels of sunlight around the world.

### 1883 Krakatau, Indonesia

The force of the eruption left a crater 290 m (951 ft) deep in the ocean floor.

### 1902 Mount Pelée, Martinique

All but two of the entire population of St Pierre were wiped out by the burning cloud of gas and dust. Around 30,000 people died.

### 1906 San Francisco, USA

Two huge tremors hit the city setting off fires that burned for many days.

### 1920 Xining, China

The entire province of Gansu was devastated by the earthquake, which killed more than 180,000 people.

### 1923 Tokyo, Japan

An earthquake of 8.3 on the Richter scale flattened 600,000 homes and knocked over stoves, which started a terrible firestorm.

A watch stopped by the eruption of Mount Pelée, 1902

### 1943 Paricutin, Mexico

Lava flowed from a crack that appeared in a farmer's field. By 1952, the cone stood 528 m (1,732 ft) tall.

### 1963 Surtsey, Iceland

Undersea volcanic explosions created a new island off the southwest coast of Iceland.

### 1963 Skopje, Yugoslavia

A violent earthquake left three-quarters of the town of Skopje homeless. At least 1,000 people were killed.

The ruins of San Francisco, 1906

Cinders engulf Vestmannaeyjar, Iceland

### 1973 Heimaey, Iceland

Eldfell volcano erupted after 5,000 years of dormancy. Molten lava engulfed one-third of the town of Vestmannaeyjar on the Icelandic island of Heimaey.

### 1976 Tangshan, China

The most disastrous earthquake in modern times. A tremor of 8.3 on the Richter scale killed over 240,000 people.

### 1980 Mount St Helens, USA

The eruption of Mount St Helens devastated vast areas around the volcano.

### 1985 Mexico City, Mexico

Powerful tremors, measuring 8.1 on the Richter scale, shook Mexico City for three minutes. One million people were left homeless.

### 1985 Nevado del Ruiz, Colombia

The eruption of the Ruiz volcano caused a massive mudflow that engulfed the town of Armero 60 km (37 miles) away.

### 1991 Kilauea, Hawaii

Kilauea volcano suddenly produced large quantities of lava that buried 13 km (8 miles) of road and 181 homes.

### 1991 Pinatubo, Philippines

The most violent volcanic eruption of the twentieth century. Around 42,000 homes destroyed.

### 1994 Los Angeles, USA

The earthquake destroyed nine highways and 11,000 buildings in 30 seconds.

### 1995 Kobe, Japan

The most powerful earthquake to hit a modern city, measuring 7.2 on the Richter scale.

### 1996 Grimsvotn, Iceland

A 4-km- (2.5-mile-) long fissure appeared in the side of the Grimsvotn volcano. Lava melting the glacier ice also damaged roads, pipelines, and power cables.

### 1998 New Guinea

A violent offshore earthquake caused a 10-m- (33-ft-) high tsunami, which swept 2 km (1.2 miles) inland. Around 4,500 people died.

### 2002 Democratic Republic of Congo

About half a million people were forced to leave their homes when rivers of lava flowed from Mount Nyiragongo. The lava destroyed two-fifths of the town.

### 2004 Sumatra-Andaman, Indian Ocean

An undersea earthquake triggered a series of tsunamis, killing around 187,000 people.

The destruction of the expressway, Kobe, Japan

### 2005 Pakistan

Also known as the Kashmir earthquake, this violent quake, measuring 7.6 on the Richter scale, killed around 73,000 people and left 33 million homeless.

### 2010 Haiti

A devastating 7.3 magnitude earthquake struck Haiti in January 2010. More than 220,000 people died.

### 2010 Iceland

In March 2010, Eyjafjallajökull volcano in Iceland erupted and disrupted air traffic across Europe for several weeks. About 45,000 passengers were affected.

### 2011 Japan

A 9.0 magnitude earthquake and tsunami struck Japan in March 2011. About 18,000 people lost their lives.

Eyjafjallajökull volcano, Iceland

# Find out more

Volcanoes are unpredictable and can be extremely dangerous. However, there are volcanic parks all over the world where you can see volcanic features safely. There are also websites where you can find news of the latest eruptions and even watch volcanic activity live.

### Yellowstone Park, USA

These tourists are walking through the Norris geyser basin in Yellowstone National Park, Wyoming, USA. Yellowstone lies on a volcanic hot spot and contains volcanic features such as geysers and fumaroles.

### The story of an eruption

The remains of the cities of Pompeii and Herculaneum in Italy (pp. 26–32) give visitors an insight into both volcanoes and life in Roman times. Both cities were destroyed when Vesuvius erupted in 79 CE, but the volcano is still active – it last erupted in 1944. Visitors can climb up and look into the crater of the volcano.

*Bodies of ancient Romans suffocated by the poisonous gases were preserved in the volcanic ash*

The ruins of Pompeii lying in the shadow of Vesuvius

## PLACES TO VISIT

**NATURAL HISTORY MUSEUM, LONDON**
- The Earth galleries have displays on volcanoes and earthquakes.

**ETNA, VESUVIUS, AND STROMBOLI, ITALY**
- These volcanoes can all be climbed with guides, weather and conditions permitting.

**THE CANARY ISLANDS**
- You can visit the volcano on La Palma and the volcanic landscape of Lanzerote.

**ICELAND**
- Iceland has over 20 volcanoes, but is most famous for its geysers and hot springs.

*Runny pahoehoe lava flowing over a cliff made up of layers of lava*

### Visiting volcanoes

There are national parks worldwide where visitors can see signs of volcanic activity, past and present. In Hawaii, visitors can drive to the rim of the active volcano, Kilauea, walk through a lava tube, and see an eruption from a safe distance.

Lava flow from Kilauea volcano in Hawaii

# WATCHING THE EARTH

In the box below you will find the addresses of some websites with general background information on volcanology and links to lots of specific volcanoes. The Smithsonian Institution in the USA produces an online weekly report on volcanic activity worldwide.

## Living with a volcano

These children are learning about Sakurajima volcano in Japan. It is one of the most active volcanoes in the world, erupting around 150 times a year. Sakurajima is monitored by a volcano observatory which collects data that helps predict volcanic activity.

## Watching from space

This picture, taken by a Space Shuttle, shows thick clouds of ash and dust from the eruption of Kliuchevskoi volcano in Russia, in 1994. Volcanic ash clouds can cause pollution and affect climate. They are watched and measured by satellites.

### USEFUL WEBSITES

- A general introduction to volcanoes:
  **www.geo.mtu.edu/volcanoes**

- Smithsonian Institution report on volcanic activity:
  **www.volcano.si.edu**

- List of top ten earthquakes and volcanoes:
  **www.uwgb.edu/dutchs/PLATETEC/TOPTEN.HTM**

- On how to build a volcano model:
  **www.sciencebob.com/experiments/volcano.php**

## Dangerous work

This volcanologist is checking gas samples inside the crater of Mount Erebus, Antarctica. It is dangerous work and several volcanologists have been killed by unexpected eruptions. There are lots of sites on the Internet where you can learn more about a volcanologist's work and find out how to become one.

# Glossary

**AA** Hawaiian word to describe thick lava that forms angular lumps when cool.

**AFTERSHOCKS** Smaller earth tremors that happen after an earthquake. These may occur weeks after the main tremor.

**ASH AND DUST** The smallest fragments of lava formed when a volcano explodes. Small pieces are called ash, and the powder-fine particles are known as dust.

**BASALT** Dark, fine-grained volcanic rock, formed by runny lavas. Basalt is the most common volcanic rock.

**BLACK SMOKER** A volcanic hot spring on the ocean floor that spits out black water rich in minerals.

**BOMBS AND BLOCKS** Large pieces of lava thrown out during a volcanic eruption. Bombs are slightly rounded, while blocks are more angular.

**CALDERA** A giant, bowl-shaped crater at the top of a volcano, formed when the summit collapses into the volcano's magma chamber.

**CARBONIZE** To turn to carbon. Objects that contain carbon will turn to carbon (or charcoal), rather than burn, when there is not enough oxygen available for them to burn in the usual way.

Carbonized walnuts from Pompeii

The crater of Mount Vesuvius, Italy

A red-hot flow of aa lava, Hawaii

**CORE** The centre of the Earth, made up of dense metals, in particular, iron. The inner core is solid while the outer core is liquid metal.

**CRATER** Hollow dip formed when the cone of a volcano collapses inwards.

**CRATER LAKE** Lake formed when water fills the crater or caldera of a volcano. Also known as lava lake.

**DORMANT** The term used to describe a volcano that has not been active for more than several hundred years but was active within the last several thousand years.

**EPICENTRE** The point on the Earth's surface, directly above the focus, or point of origin, of an earthquake.

**EXTINCT** The term used to describe a volcano that has not been active for more than several thousand years.

**FAULT** A fracture in rock where blocks of rock slide past each other.

**FEEDER PIPE** The long tube through which magma passes to reach the surface.

**FISSURE** A crack in the earth. A fissure eruption is one where runny lava flows from the crack.

**FOCUS** The point within the Earth's mantle from which an earthquake originates.

**FUMAROLE** A vent or opening in the Earth's surface that releases steam or gas.

A geyser in Iceland

**GEOLOGY** The study of the history and development of Earth's rocks. The people who study geology are geologists.

**GEYSER** A place where water that has been superheated by hot magma bursts up into the air.

**HOT SPOT** A place in the middle of a tectonic plate, rather than at the boundary, where columns of magma from the mantle rise up through the crust creating a volcano.

**HYDROTHERMAL VENT** A place where mineral-rich water heated by hot magma underground erupts onto the surface. Geysers, black smokers, and hot springs are all hydrothermal vents.

**IGNEOUS ROCKS** Rocks formed as hot magma and lava cools.

**INTENSITY** The term used to describe the severity of the shaking experienced during an earthquake. It is usually measured using the Modified Mercalli Intensity Scale.

**LAPILLI** Small fragments of lava, formed as the magma bursts out of a volcano.

**LAVA** Hot, molten rock that erupts from a volcano.

**LAVA TUBE** A tunnel of lava created when the surface of a lava flow cools and hardens to form a roof, while hot, molten lava continues to flow inside it.

**MAGMA** Hot, molten rock originating from within Earth.

**MAGMA CHAMBER** Area beneath a volcano where magma builds up before an eruption.

**MAGNITUDE** The term used to describe the severity or scale of an earthquake. This is measured in several ways, the most common of which is the Richter scale.

Aa lava

**MANTLE** The layer between the Earth's crust and its core. The mantle is 2,900 km (1,800 miles) thick.

Pahoehoe lava

**MERCALLI SCALE** The scale for measuring the intensity of an earthquake by observing its effects.

**MID-OCEAN RIDGE** A mountain ridge on the ocean floor formed where two tectonic plates meet.

**MUDFLOW** A fast-moving stream of mud, water, and often volcanic ash and pumice.

**PAHOEHOE** The Hawaiian term for hot, runny lava that flows quickly and, usually, in quite shallow flows.

**PILLOW LAVA** Rounded lava formations shaped when lava erupts gently underwater.

**PLATE TECTONICS** The theory that the Earth's surface is broken up into large slabs or plates that are moving constantly at the rate of a few centimetres a year. Most volcanoes and earthquakes are found at the boundaries of these plates.

**PUMICE** Lightweight volcanic rock filled with holes formed by the bubbles of gas in the lava.

Pumice

**SEISMIC WAVES** The vibrations, or shock waves, that radiate out from the focus of an earthquake.

**SEISMOGRAM** The record produced by a seismograph, showing the pattern of shock waves from an earthquake.

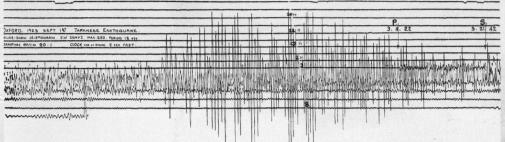
Seismogram of the 1923 Tokyo earthquake

**PYROCLASTIC FLOW** A burning cloud of gas, dust, ash, rocks, and bombs that flows down the mountain after an explosive eruption. If the cloud is made up of more gas than ash, it is known as a pyroclastic surge.

**P WAVES** The fastest and first or "primary" waves of an earthquake.

**RICHTER SCALE** Scale for measuring the total energy released by earthquakes and recorded by seismographs. The scale ranges from 1 to 10, with 10 at the most severe end of the scale.

**RING OF FIRE** The area encircling the Pacific Ocean where most volcanic and earthquake activity occurs.

**SEISMOMETER** A machine for detecting earthquake shock waves. The machine that records the data is a seismograph.

**SUBDUCTION ZONE** The area where two tectonic plates meet and one plate is pushed down into the mantle, partly melting rocks. The resulting magma erupts at the surface through volcanoes.

**S WAVES** The slower, "secondary" waves of an earthquake.

**TSUNAMI** Fast-moving waves caused by earthquakes or volcanic eruption that displace the ocean floor and water.

**VENT** The opening through which a volcanic eruption occurs.

**VOLCANOLOGIST** A scientist who studies volcanoes.

Volcanologists collecting gas samples on Colima volcano, in Mexico

# Index

# Acknowledgements

**Dorling Kindersley would like to thank:**
John Lepine & Jane Insley of the Science Museum, London; Robert Symes, Colin Keates & Tim Parmenter of the Natural History Museum, London; the staff at the Museo Archeologico di Napoli; Giuseppe Luongo; Luigi Iadicicco & Vincenzo D' Errico at the Vesuvius Observatory for help in photographing the instruments on pp. 49, 53 & 55; Paul Arthur; Paul Cole; Lina Ferrante at Pompeii; Dott. Angarano at Solfatara; Carlo Illario at Herculaneum; Roger Musson of the British Geological Survey; Joe Cann; Tina Chambers for extra photography; Gin von Noorden & Helena Spiteri for editorial assistance; Céline Carez for research & development; Wilfred Wood & Earl Neish for design assistance; Jane Parker for the index; Stewart J. Wild for proof-reading; Carol Davis for art-editing the jacket; Neville Graham, Sue Nicholson, & Susan St. Louis for the wallchart.

**Illustrations:** John Woodcock

**Maps:** Sallie Alane Reason

**Models:** David Donkin (pp. 8–9, 50–51) & Edward Laurence Associates (pp. 12–13)

For this relaunch edition, the publisher would also like to thank: Hazel Beynon for text editing, and Carron Brown for proofreading.

The Publishers would like to thank the following for their kind permission to reproduce their photographs

a-above; b-below; c-centre; l-left; r-right; t-top
Ancient Art & Architecture Collection: 47t.
Art Directors Photo Library: 47tc, 49tc.
B.F.I.: 46cr, 57cr. Bridgeman Art Library: 6tl & c.
Musee des Beaux-Arts, Lille: 44tl. British
Museum: 27tr & c. Herge/Casterman: 20tl. Dr
Joe Cann, University of Leeds: 25bc. Jean-Loup
Charmet: 27bl, 31tl, 46bl, 51tl. Circus World
Museum, Baraboo, Wisconsin: 32bl. Corbis:
Bettmann 67tl; Danny Lehman 65cr; Vittoriano
Rastelli 70tc; Roger Ressmeyer 71br; Michael S.
Yamashita / National Geographic Society (57cl).
Eric Crichton: 41tr, 41bl. Culver Pictures Inc:
60br. DK Images: Satellite Imagemap © 1996–
2003 Planetary Visions 9c. Dreamstime.com:
Photographerlondon (67b). Earthquake
Research Institute, University of Tokyo: 63cr.
Edimedia/Russian Museum, Leningrad: 28cl.
E.T. Archive: 49c, 58tr, 62tl. Mary Evans Picture
Library: 8tc, 16tl, 26tl, 27br, 28c, 29cr, 46tl, 64cr,
66br. Le Figaro Magazine/Philippe Bourseiller:
19t, 19cl 19bl, 19br, 35tr. Fiorepress: 49br.
Gallimard: 44tr. G.S.F.: 13tl, 14bl; /Frank Fitch:
20c. John Guest c.NASA: 44bl. Robert Harding
Picture Library: 12tl, 14tr, 15cr 17tr, 17cr, 20tr,
21br, 63bl, 34cr, 38tl, 38br, 39br, 40tr, 42c, 43br, 48br,
49bl, 51tr, 60lc, 62bl, 62–63c, Explorer 66bl,
Christian Kober (59bl). Bruce C. Heezen & Marie
Tharp, 1977/c.Marie Tharp: 11c. Historical
Pictures Service, Inc.: 10cr. Michael Holford:
22tr. Illustrated London News: 60bl. ImageState:
68bc. Katz Pictures: Alberto Garcia/Saba 64b.
Frank Lane Picture Agency: 23rct, 23c. Frank
Lane Picture Agency/S.Jonasson: 41tc, 49cr.
Archive Larousse-Giraudon: 32cl. London Fire
Brigade/LFCDA: 58bl. Mansell Collection: 31tr.
NASA: 24bc, 55cr, 68–69. Natural History
Museum: 34tl. National Maritime Museum: 8tl.
Orion Press: 7tr, 61bl. Oxford Scientific Films/
Colin Monteath: 69br; /NASA: 69cl; /Kim
Westerkov: 11bl; 15tl. Planet Earth Pictures/
Franz J.Camenzind: 7cr; /D.Weisel: 17br; /James
D. Watt: 23rcb; /Robert Hessler: 25tr, 25lc.
Popperfoto: 6b, 11tr, 54tl. R.C.S. Rizzoli: 48lc.
Rex Features: 67tr. Gary Rosenquist: 14tl, 14br,
15tr, 15br. Scala: 7c, 49bl(inset); /Louvre: 63cr.
Science Photo Library/Earth Sattelite Corp.: 7tl;
/Peter Menzel: 7br, 20br, 46cl, 65bl; /David
Parker: 13tr, 61tl, 61tr, 65tc; /Ray Fairbanks: 18c;
/Inst Oceanographic Sciences: 24c; /Matthew
Shipp: 24bl; /NASA: 35c, 43cr, 44cr, 44c, 44–45b;
/U.S.G.S.: 45tr; /Peter Ryan: 55tr; /David

Weintraub 66t. **Frank Spooner Pictures:** 8cl,
16cr, 17tl, 22bl, 23tr, 23br, 35tc, 42cl, 47cl, 49cl,
56tl, 56c, 56bl, 57bl, 58tl, 59tr, 59br, 61lcb; /Nigel
Hicks 69tr. **Syndication International:** 31cr; /Inst.
Geological Science: 32tc, 35br, 41tl, 47br; /Daily
Mirror: 54tr, 57tr. **Susanna van Rose:** 39bl.
Woods Hole Oceanographic Institute/Rod
Catanach: 24br; /Dudley Foster: 25bl; /J.
Frederick Grassle: 25tc; /Robert Hessler: 25tl.
**ZEFA:** 9tl, 21t, 34bl, 45br, 57br.

**Wallchart: Corbis:** Burstein Collection / Barney
Burstein fbr; Roger Ressmeyer cra; **DK Images:**
Museo Archeologico Nazionale di Napoli cr;
**Natural History Museum,** London clb (aa lava),
cra (pyroclastic debris); **Getty Images:** Image
Bank / G. Brad Lewis cla (Pahoehoe flow);
**National Geographic / Klaus Nigge** ca; **NASA:**
crb; **Science Photo Library:** Peter Menzel br
(San Francisco)

All other images © Dorling Kindersley.
For further information see:
www.dkimages.com